Mfon Obot
Danjuma Yawas

Development of Abrasive Sandpaper using Periwinkle Shell

Mfon Obot
Danjuma Yawas

Development of Abrasive Sandpaper using Periwinkle Shell

LAP LAMBERT Academic Publishing

Imprint

Any brand names and product names mentioned in this book are subject to trademark, brand or patent protection and are trademarks or registered trademarks of their respective holders. The use of brand names, product names, common names, trade names, product descriptions etc. even without a particular marking in this work is in no way to be construed to mean that such names may be regarded as unrestricted in respect of trademark and brand protection legislation and could thus be used by anyone.

Cover image: www.ingimage.com

Publisher:
LAP LAMBERT Academic Publishing
is a trademark of
Dodo Books Indian Ocean Ltd. and OmniScriptum S.R.L publishing group

120 High Road, East Finchley, London, N2 9ED, United Kingdom
Str. Armeneasca 28/1, office 1, Chisinau MD-2012, Republic of Moldova, Europe
Printed at: see last page
ISBN: 978-3-659-80203-4

ACKNOWLEDGEMENT

I first of all thank the Almighty God who made this possible and for His abiding grace, mercy and strength granted to me to fulfill my degree on Master of Science. To Him be all praise. I also acknowledge my supervisory team of Dr. D. S. Yawas and Prof S. Y. Aku for their kind guidance and direction in the course of my work.

Without reservations I express thanks to Dr. Asuke, Dr. Aigbodion, Engr. Amaren, Mr. Danladi, Peter of woodwork dept., whose inspirations and support went a long way to ordering my progress in the work.

My gratitude also extends to my colleagues and friends Bello, Iorpenda, Julius, Aderemi and Adegoke to mention a few for their support, criticism and encouragement. To my family made up of my ever accommodating and supporting mother and my siblings, I owe you all my love.

DEDICATION

This research work is dedicated to my family, Barr (Mrs) Ime Obot, Sylvia, Laura and Bridget Obot, and to the glory of the Most High God who is the custodian of all grace.

TABLE OF CONTENTS

LIST OF FIGURES

LIST OF TABLES

LIST OF PLATES

LIST OF ABBREVIATIONS AND SYMBOLS

PWS -   Periwinkle Shell

PKS -   Palm Kernel Shell

SEM -   Scanning Electron Microscope

XRF -   X-ray Florescent Test

XRD -   X-ray diffractometry

EDS -   Energy Dispersive Spectroscopy

OES -   Optical Emission Spectrophotometry

% -   Percentage

^{o}C -   Degree Celcius

g -   Gram.

Sec -   Seconds

N/mm^{2} -   Newton per square millimeters

mm -   Millimeters

wt -   Weight

μm -   Micrometer

CHAPTER ONE

INTRODUCTION

1.1   Background of the Study

The Nigerian manufacturing sector is described as weak and its contribution to growth is suboptimal (Annual Performance Report, 2011). Nigeria has been riddled with challenges among which are import dependence, neglect of manufacturing due to the nation's reliance on crude oil revenue and bias against indigenous products in the Nigerian populace.

At the turn of Nigeria's democracy in 1999, successive governments have made efforts to improve the situation in the manufacturing sector and reposition it in becoming a major contributive block to economic development and ultimately the GDP of the nation. Such ameliorative measures include the creation of the Local Content Act which focuses on promoting value addition to the Nigerian economy through the utilization of local raw materials, products and services in order to stimulate growth of indigenous capacity (Omobowale, 2010).

Efforts have been made in the past by different researchers to utilize locally sourced materials in the formulation and development, manufacture and construction of products such as building materials, design and manufacturing tools and equipment etc. Researches in Nigeria and all over the world today are focusing on ways of utilizing either industrial or agricultural wastes as a source of raw materials in industry. Utilization of these wastes will not only be economical, but may also result in foreign exchange earnings and environmental controls. (Aku et al, 2012)

The unsustainability of imported goods to the economy of the nation as well as the need for environmental waste management call for more investment in locally sourced raw materials and biodegradable recyclable wastes as raw material substitutes in producing tools for industrial and domestic use. This work seeks to address the recycling of agricultural wastes for producing tools used for abrasive machining processes.

1.2. Statement of the Problem

This work considers firstly the evolving government policies towards importation of foreign products which can be made locally and the need to invest in locally sourced materials as raw materials in manufacture of commonly used industrial tools.

In addition, importing foreign abrasive tools into the Nigerian market is expensive to the Nigerian economy, and to the end user. The high cost in imported products could be attributed to price fluctuation as a result of low exchange rate, attendant imported inflation, landing and hauling costs, overhead charges, and corruption in the system (Benjamin, 1989).

Finally, the locally sourced materials to be considered such as periwinkle and palm kernel shells are inclusive of environmental wastes. Periwinkle shells incur environmental pollution particularly in the southern and riverine areas of the country were this impact is felt most. Palm kernel shells are recovered as by-products in palm oil production. Large quantities of these shells are generated annually and only some fractions are used for applications such as palliatives for un-tarred roads and in producing activated carbon. The unused shells are dumped around the processing mill, constituting environmental and economic liability for the mill (Koya and Fono, 2009). This study aimed at waste- recycling of periwinkle and palm kernel shells.

1.3. The Present Study

This present work aims at evaluating the feasibility of using periwinkle and palm kernel shell recyclable wastes that will be processed into abrasive particles called grits used for polishing metallic surfaces and for sanding uneven wood surfaces. The periwinkle grains and palm kernel shell grains will be held by polyester resin bond unto the surface of fabric and paper used for the study.

1.4.   Aim and Objectives

The aim of the study is to produce emery cloth and sand paper by the hand spray method (Wai and Lilly, 2002) using processed periwinkle and palm kernel shell grains as grits and polyester resin as bond, to characterize the bond mixture for each of them and to determine their effectiveness in abrading operations.

The specific objectives of the study are:

i. To carry out physico-chemical characterization of periwinkle and palm kernel shells.

ii. To process the periwinkle and palm kernel shells by crushing and sieving into varied particle sizes of 105μm, 250μm and 425μm which are P140, P60 and P40 sandpaper grit sizes (according to Federation of European Producers of Abrasive standard).

iii. To determine binding properties of polyester resin on the periwinkle and palm kernel shell grains by varying the weight percent of resin from 4 – 12 weight percent with fixed percentage of catalyst and accelerator (0.5 weight % each).

iv. To determine the physical and mechanical tests of periwinkle grains/polyester resin, palm kernel shell grains/polyester resin composites such as density, percent adsorption, compressive strength, hardness and wear.

v. To use composition with best properties in producing emery cloth and sand paper samples using hand spraying method (Wai and Lilly, 2002) into medium (P40 and P60) and fine (P140) grade of emery cloth and sandpapers.

1.5.   Justification of the Study

In order to be effective as abrasive grits, the periwinkle and palm kernel shell grains should have properties such as hardness, resistance to attritious wear, brittleness, friability and chemical stability.

From previous researches done by Aku et al., (2012), Dagwa et al., (2012) among others, periwinkle and palm kernel shell grains have both been used as filler

3

reinforcements in polymer matrix composites for load bearing, wear resistance and hardness applications for example in brake pad production, and they have been proven successful in these applications.

The study will go further to look into the wear and abrasive properties of periwinkle and palm kernel shell grains in polyester resin matrix bond in development of the abrasive materials.

1.6. Significance of the Study

1. The study will provide a way of converting agro resources of recyclable nature into a useful engineering tool such as emery cloth and sand paper.

2. The study keys into the Federal Government mandate of enhancing local content in rendering produce and services to the country and at the same time curtailing dependence on importation.

3. The production method adopted for the study (hand spray method) shall provide a cheap alternative with low cost of capital to the conventional methods of producing emery cloth and sand paper which is by electrostatic method or mechanical method.

4. The study which is developing a product manufacturing process shall provide a means of livelihood for entrepreneurs and small business owners in the south-south region where periwinkle and palm kernel shells abound.

1.7. Scope of the Study

The study is to develop emery cloth and sand paper using periwinkle and palm kernel shells which are agricultural wastes in southern Nigeria and polyester resin as bond material by the hand spray method. The process will be carried out at ambient temperature and will not involve electrostatic process or sintering as in established methods.

CHAPTER TWO

LITERATURE REVIEW

2.1. Introduction

This chapter outlines relevant journal reports and textbook literature on periwinkle shells, palm kernel shells, abrasive machining and polyester resins.

2.2. Applications of Periwinkle Shells

Periwinkles (*Nodilittorina radiata*) are small edible species of medium-sized sea snails characterized by their black to grey colored, hard spiral and conical shaped outer shells with sizes ranging from 1.3 to 2.5 cm in length. The two species of periwinkle commonly found in the estuarine habitat of the Niger Delta are Tympanotonus fuscatus and Pachymelania aurita (Bob-Manuel, 2012). The fresh periwinkle is edible and also used as bait by fisher folks. They are rich in protein (about 21%), vitamins and minerals (Egonmwan 1980). Akwa Ibom delicacies such as Ekpan Nkukwo, Afang and Atama Soup, are rarely served without periwinkle. They are considered as cheap but nevertheless vital food source. Usually the periwinkle is prepared in two ways; either by washing and cooking the periwinkle meat in together with the shell, or discarding the shell while the meat inside is used for dishes. In a cultural sense, the shells are known to have very little value to the people. This can be attributed to the commonness of the shells, the foul smell they emit, and its unpleasant feel (due to the hard abrasive projections from the shell) as well as their unattractive color which is predominantly black. As a result, they are mainly considered as useless waste products. With all these limitations, yet a fraction of them is used in creative ways to serve mankind. Some locals have considered them useful as scouring powder for sooty kitchen pots after being ground into powder. Others have used the shells as earth fillers in making roads as substitute for gravel. They have also found use in local arts and crafts such as bead-making etc. As regarding local use as scouring powder, the major setback to this is the need of energy to crush the shells into powder.

5

Studies have been carried out in the past on other benefits that can be harnessed from the periwinkle shells aside their common uses. Agoha (2007) produced useful biomaterials such as chitin and chitosan from waste periwinkle shells. Malu and Bassey (2003) evaluated the suitability of periwinkle shells as a substitute for lime in glass manufacturing. They did a proximate analysis of periwinkle shell, which showed that the shell contained important minerals suitable for glass production such as calcium oxide (38.4%), silicon IV oxide (0.014%), magnesium oxide (18.70%), aluminium trioxide (0.211%) and iron oxide (0.019%). Oribhabor and Ansa (2006) considered the use of periwinkle for the formulation of fish feed as a source of calcium. Currently periwinkle shells are used as sources of calcium essential in the diet of poultry layers, which utilizes the mineral in forming the shells of their eggs. Periwinkles have been used for the production of adsorbents for the treatment of lead in industrial waste water (Badmus et al., 2007).

In engineering applications, these seashells are being used as substitutes for aggregates (chippings) especially in the coastal areas where aggregates are lacking. Some researchers have conducted exploratory studies on the partial or total substitution of waste sea shells with coarse aggregates for the production of mortar and concrete used for civil construction. In the study carried out by Adewuyi and Adegoke (2008), they found that the replacement of granite with 35.4 -42.5% waste periwinkle shells did not compromise the compressive strength of the resulting concrete and made 14.8 to 17.5% saving in material cost. Osarenwinda and Awaro (2009) carried out geotechnical analysis of concrete produced from periwinkle waste shells and found that the periwinkle shell had a bulk density of 517 kg m^{-3} and specific gravity of 2.05. Their results also showed that a design mixture of 1:1:2, 1:2:3 and 1:2:4 of cement: sharp sand: periwinkle shell ratios have a compressive strength of 25.67, 19.50 and 19.83 Nmm^{-2}, respectively at 28 days hydration period, which met the ASTM -77 recommended standard minimum strength of 17 Nmm^{-2} for structural light weight concrete. Studies into the use of periwinkle as composite materials have also been carried out. It has been seen that periwinkle shells are mainly selected for applications requiring wear resistance and hardness properties. This without surprise is as a result of periwinkle shells being mainly ceramics which are ideal for these applications due to their strong covalent/ionic bonds.

2.3. Applications of Palm Kernel Shells

Palm kernel shells are the crushed outer part of palm kernel nut derived after the extraction of palm oil. It is obtained as crushed pieces after threshing or crushing to remove the seed which is used in the production of palm kernel oil (Olutoge et al., 2012). Palm kernel shells are available in large quantities in palm oil producing areas of southern and Middle belt states of Nigeria such as Akwa Ibom, Rivers, Imo, Edo, Osun and Kogi states.

Many varieties of oil palm exist, which include dura, pisifera and tenera and they are recognized mainly by the thickness of their shell (endocarp) and fibrous oily part (mesocarp) and the fruits. The dura variety has very thick shell and thin fibrous part only. In the pisifera variety, the shell is almost absent or very tiny, the bulk of the fruit being fibrous with little or no kernel. The tenera variety is the hybrid of dura and pisifera. The thickness of the shell and the fibrous part are of medium size (Usman et al., 2012).

Palm kernel shells in local applications are used as a good source of fuel for domestic cooking in most areas where they are produced (Usman et al., 2012). Palm kernel shells are used in some areas for making terrazzo; they are also used as fill materials for filling pot holes in muddy areas in some localities.

In engineering applications, palm kernel shells have been used as filler reinforcements in matrices of different types of materials such as concrete, polymers and as material substitutes for given applications requiring friction and wear resistance. Olutage et al, (2012) investigated the use of saw dust and palm kernel shells as replacement for fine and coarse aggregates in reinforced concrete slabs. Koya and Fono (2009) produced palm kernel shell based brake pads as substitutes for asbestos which has been found to be carcinogenic. Ibhadode and Dagwa (2008)

7

worked on development of asbestos-free friction lining material from palm kernel shell. Dagwa et al, (2012) characterized palm kernel shell powder for use in polymer matrix composites. They carried out tests on the powder such as powder porosity, hydration capacity, moisture sorption, particle size distribution, bulk density, tapped density, powder flow, pH of powder dispersion as well as differential scanning calorimetry thermal (DSC), X-ray diffraction (X-RD) and scanning electron microscopy (SEM). The EDX analysis presents the elemental weight percent (wt. %) obtained from raw palm kernel shell, and showed the presence of carbon, oxygen, aluminium, silicon, phosphorous, and potassium with 63.02, 36.04, 0.43, 0.17, 0.17, and 0.17wt %, respectively.

2.4. Abrasive Machining Process

Abrasives are defined as small hard particles having sharp edges and irregular shapes (Kalpakjian, 2006). Abrasive machining is the process of material removal by the action of abrasive particles usually attached to either a flexible backing material or a rigid core or base. It can also be referred to as grinding. To grind means 'to abrade', to wear away by 'friction' or 'to sharpen' (Jain, 2008). In grinding, the material is removed by means of a rotating abrasive wheel. Abrasive process also involves various techniques such as honing, lapping, polishing and abrasive jet machining.

The abrasive machining process has the following advantages:

1. It is the convenient method of finishing process after all the rough finishing and heat treatment operations have been carried out.
2. It is an important approach to precision finishing and ultra-precision machining; it can easily produce surfaces with high precision and low roughness values.
3. Various high-hardness materials can be machined, especially hard and brittle materials such as optical glass, ceramics and semiconducting materials.
4. It is widely applied to various industries such as mechanical manufacturing, woodwork industry, construction and refractory, due to its high technical adaptability unto various materials and surfaces.

5. It has been increasingly applied to mass production where high quality stability is required.

2.4.1. Grinding

Grinding is the most important abrasive machining process (Jain, 2008). Grinding wheels are made of natural or synthetic abrasive minerals bonded together in a matrix to form a wheel. Grinding can be used on all types of materials as a finishing operation achieving surface finishes up to 0.025 μm and extremely close tolerances.

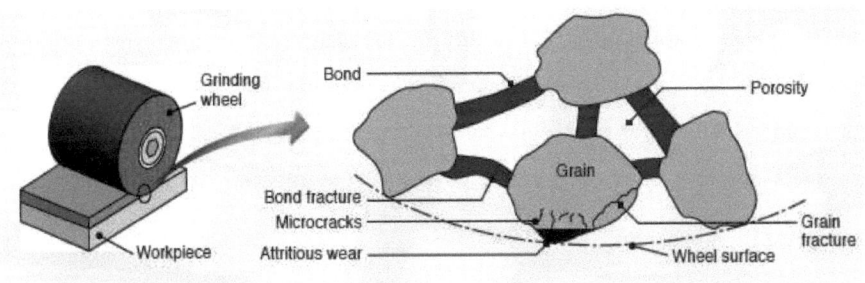

Fig 2.I: Physical model of grinding wheel showing structure (Kalpakjian and Schmid, 2006)

2.4.2. Polishing

Polishing is an abrasive machining procedure that can produce a smooth, lustrous surface finish on metal surfaces. The lustrous effect is as a result of softening and smearing of surface layers (Kalpakjian, 2006).

Fig 2.2: Sandpaper surface showing abrasive grains (arcabrasives.com, 2013)

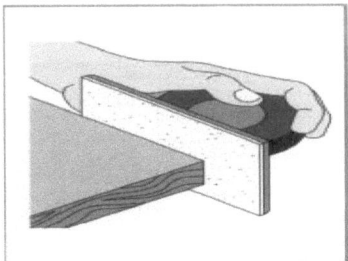

Fig 2.3: Manual Sanding of wood to smoothen surface (arcabrasives.com, 2013)

Sand papers and emery clothes sometimes referred to as *flexible abrasives* or *coated abrasives* are articles used for manual or machine smoothing, polishing, cleaning and finishing of wooden, metallic or glass materials (see Fig 2.2-2.3). They consist of a flexible backing material on which a single layer of glued abrasive grit is sprayed. The glue or adhesive used to bond the abrasive to the backing could be all resin, which is waterproof, glue or a combination of the two.

2.5. Type of Abrasives

The particular abrasive is chosen based on the way it will interact with the work material. The ideal abrasive has the ability to stay sharp with minimal point dulling. When dulling begins, the abrasive fractures, creating new cutting points. The mechanism of failure for each abrasive type is unique with distinct properties for hardness, strength, fracture toughness and resistance to impact (Pruti, 2011).

2.5.1. Grinding abrasives

Kalpakjian and Schmid (2006) categorized abrasives for grinding into conventional and super abrasives as shown in table 2.1. Under the categories of conventional abrasives are aluminium oxide (Al_2O_3) and silicon carbide (SiC). The super abrasives are the hardest materials known to man, and they are Cubic boron nitride (CBN) and Diamond.

Table 2.1: Ranges of Knoop Hardness for Various Materials and Abrasives (Kalpakjian and Schmid, 2006)

Common glass	350 -500	Titanium nitride	2000
Flint, quartz	800 -1100	Titanium carbide	1800 -3200
Zirconium oxide	1000	Silicon carbide	2100 -3000
Hardened steels	700 -1300	Boron carbide	2800
Tungsten carbide	1800 -2400	Cubic boron nitride	4000 -5000
Aluminium oxide	2000 -3000	Diamond	7000 -8000

The type of material to be ground affects the selection of abrasive, grain size and grade. Alumina abrasives (UGWC 1992) are used for grinding high tensile materials such as steel and ferrite cast irons. Silicon carbide abrasives that are even more friable are used for grinding low tensile strength materials and non-metallic materials. CBN abrasive wheels (King and Hahn, 1986) are suitable for grinding high speed steel and high alloy steels. Carbides, ceramics, glass and plastics are often ground using diamond wheels. The harder the workpiece, the harder the grain required. For particular grain hardness, a hard workpiece requires a softer bond than a soft workpiece (UGWC 1992).

– **Natural Abrasives** are commonly found in nature. These are emery, corundum, quartz, garnet and diamond. Due to the contents of impurities and their non-uniform properties, their performance is inconsistent and unreliable and hence the need for man-made abrasives produced under controlled conditions.

– **Aluminium Oxide** is the most common abrasive used in grinding wheels (Yamaguchi et al., 1999). There are many different types of aluminum oxide abrasives, each specially made and blended for particular types of grinding jobs. Each abrasive type carries its own designation (usually a combination of a letter and a number). These designations do vary from one manufacturer to another. Aluminum oxides are chosen for grinding carbon steel, alloy steel, high speed steel, annealed malleable iron, wrought iron, and bronzes and similar metals. Aluminum oxide crystals are used principally in grinding ferrous and other materials that have a high tensile strength because they are tough and resist fracture to a high degree.

– **Silicon Carbide** is harder than aluminum oxide, but its crystals are not as tough and break easily. Silicon carbide crystals fracture easily and the material is especially adapted to cutting materials with low tensile strength such as brass, aluminum, copper, cast iron, rubber, and plastics. It is also used in grinding hard, brittle materials such as carbide, stone and ceramics (Kalpakjian and Schmid, 2006).

- **Ceramic Aluminium** is the newest major development in abrasives. This is a high purity grain manufactured in a gel sintering process. The result is an abrasive with the ability to fracture at a controlled rate at the sub-micron level, constantly creating thousands of new cutting points. This abrasive is exceptionally hard and strong. It is primarily used for precision grinding in demanding applications on steels and alloys that are most difficult to grind. The abrasive is normally blended in various percentages with other abrasives to optimize its performance for different applications and materials.

- **Zirconia Alumina** is abrasive made from different percentages of aluminum oxide and zirconium oxide. The combination results in tough durable abrasive that works well in rough grinding applications, such as cut-off operations, on a broad range of steels and steel alloys. As with aluminum oxide, there are several different types of zirconia alumina from which to choose (Kalpakjian and Schmid, 2006).

2.5.2. Polishing abrasives

- *Emery*: Emery is a dark gray, round-shaped grain which tends to polish rather than abrade a work surface. It is used for polishing and cleaning metal only.

- *Garnet*: Garnet is reddish brown in color. This natural abrasive is medium hard and relatively sharp, but not as durable as synthetic abrasives. It is used only on woods. It produces an excellent finish on wood work.

- *Silicon carbide*: Silicon carbide is the hardest and sharpest of the manufactured abrasives. Because of its extreme sharpness, this bluish-black abrasive grain permits fast stock removal and cool cut. It is used to polish cast iron, non-ferrous metals i.e. brass, aluminum and bronze; non-metals, i.e. glass, rubber, plastic and stone; final finish on wood and stainless steel, and abrasive planning particle board.

- *Aluminum oxide:* Light brown aluminum oxide is a tough, yet sharp, synthetic abrasive characterized by cool cut, long life and the ability to break down under pressure, producing new cutting edges. Aluminium oxides are applied in wood sanding and non-ferrous metal finishing.

−*Zirconia alumina:* Zirconia alumina is an ultra-tough, synthetic abrasive which provides a free, cool cut for high stock removal applications. It is tougher and sharper than aluminum oxide. It has a micro-crystalline structure which allows for controlled breakdown and self-sharpening. It is used in heavy-duty snagging and grinding of all ferrous and non-ferrous metals, abrasive planning of wood, plywood and particleboard grinding fiberglass, rubber and plastics (Kalpakjian and Schmid, 2006).

2.6. Abrasive Grain Size (Grit Size)

Grit sizes are measured by the fraction of bulk grains that can pass through a series of vibrating screens or meshes having a specified number of openings per square inch (Marinescu et al, 2004). Grain sizes relate to the screening mesh used for sizing. Mesh number is the number of wires per linear inch of the screen, through which the grains pass while being retained at the next finer size screen. The size of the grains reduces as the mesh number increases.

The two most common standards for abrasive grits are the United States CAMI (Coated Abrasive Manufacturers Institute, now part of the Unified Abrasives Manufacturers' Association) and the European FEPA (Federation of European Producers of Abrasives) "P" grade. There are two broad classifications of grit sizes which are; Macrogrits and Microgrits. The mean diameters of Macro and Micro grits according to the Federation of European Producers of Abrasives (FEPA) Standards are given in table 2.2. The particle size distribution of macro grits is determined by sieving, while the micro grits are measured by sedimentation method.

Selection of the size of grain will depend on the amount of material to be removed, the finish desired and the mechanical properties of the material to be ground. Generally the larger the grains, the faster the material will be removed. Coarse grains are adapted for cutting and sanding uneven surfaces. The extreme range of coarse grains such as mesh size 12 to 20 is used for extra deep and coarse cutting in workpieces. Medium grains are adapted for sanding marks and unevenness. Fine grains are used for fine-sanding between and after final surface treatment. Very fine

13

grains are adapted for final sanding, where there is a particular requirement for an extra smooth surface. Extreme fine grains with mesh size of over 800 up to 1200 are termed ultrafine and used for grains used to give extremely smooth surfaces.

Table 2.2: The mean diameters of Macro and Micro grits according to FEPA Standards (www.fepa- abrasives.org)

MACROGRITS		MICROGRITS	
Grit designation	Mean Diameter in µm	Grit designation	Mean grain size value in µm
P 12	1815	P 240	58.5 ± 2
P 16	1324	P 280	52.2 ± 2
P 20	1000	P 320	46.2 ± 1.5
P 24	764	P 360	40.5 ± 1.5
P 30	642	P 400	35.0 ± 1.5
P 36	538	P 500	30.2 ± 1.5
P 40	425	P 600	25.8 ± 1
P 50	336	P 800	21.8 ± 1
P 60	269	P 1000	18.3 ± 1
P 80	201	P 1200	15.3 ± 1
P 100	162	P 1500	12.6 ± 1
P 120	125	P 2000	10.3 ± 0.8
P 150	100	P 2500	8.4 ± 0.5
P 180	82		
P 220	68		

Table 2.3: Classification of Grain Size (Grit Number)

Classification	Grain Sizes						
Coarse	10	12	14	16	20	24	
Medium	30	36	40	46	54	60	
Fine	70	80	90	100	120	150	180
Very Fine	220	240	280	320	400	500	600

2.7. Type of Bond

The bond is the medium that holds the grains together in the form of a wheel, or on a belt or disk. The bond functions in the same way as a tool post and holds the grains or cutting tools in position until they become dull and are torn out and fresh grains are exposed.

2.7.1. Grinding wheel bond

Vitrified bonds, organic substances and rubber are the three principal types of bonds used in conventional grinding wheels (Malkin and Guo, 2008). Each type is capable of giving distinct characteristics to the grinding action of the wheel. The type of bond selected depends on such factors as the wheel operating speed, the type of grinding operation, the precision required and the material to be ground.

− **Vitrified or Ceramic bonds**: Most grinding wheels are made with vitrified bonds, which consist of mixtures of carefully selected clays. At the high temperatures produced in kilns where grinding wheels are made, the clays and the abrasive grain fuse into a molten glass condition. During cooling, the glass forms a span that attaches each grain to its neighbor and supports the grains while they grind.

Grinding wheels made with vitrified are very rigid, strong, and porous. They remove stock material at high rates and grind to precise requirements. They are very hard, but at the same time they are brittle like glass. The pressure of grinding breaks down vitrified bonds.

− **Organic or Resinoid bonds**: Bonds made of organic substances soften under the heat of grinding. The most common organic bond type is the resinoid bond, which is made from synthetic resin. Wheels with resinoid bonds are good choices for applications that require rapid stock removal as well as those where better finishes are needed. They are designed to operate at higher speeds, and they are often used for wheels in fabrication shops, foundries, billet shops and for saw sharpening and gumming.

− **Rubber bonds**: Another type of organic bond is rubber and wheels made with rubber bonds offer smooth grinding action. Rubber bonds are often found in wheels used where a high quality of finish is required, such as ball bearing and roller bearing races. They are also frequently used for cut-off wheels where burr and burn must be held to a minimum (Malkin and Guo, 2008).

2.7.2. Sand Paper/emery cloth bond

An adhesive bond system is required to secure the abrasive mineral to the backing. All coated abrasive products are made with a two stage bonding process. The first layer of bond applied to the backing is called the 'make' coat.

The make coat provides the adhesive base between the abrasive mineral and the backing. The second coat is the 'size' coat, which is applied over the abrasive mineral and 'make' coat to anchor the abrasive mineral and provide the desired physical strength of the finished product. An illustration of sand paper showing these components is given in figure 2.4.

Glue, urea resin, and phenolic resin are the three basic bonding agents most commonly used (US Environmental Protection Agency, 2013). There are many size coats and make coat combinations, such as glue over glue, urea over glue, and resin over resin. Glue over glue is the most flexible bond while resin over resin bond is moisture-resistant, harder, less flexible, heat-resistant and has superior grain retention.

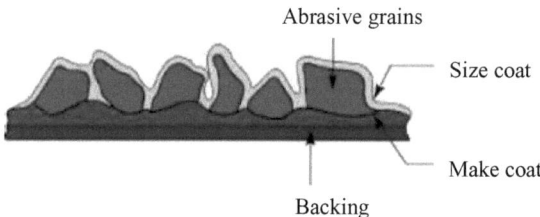

Fig 2.4: Illustration of Sand Paper showing composition (www.madehow.com)

2.8. Coated Abrasive Backing Types

Backings are the base for the abrasive minerals and, combined with the adhesive bond, support and anchor the abrasive mineral. The backings used in the manufacture of coated abrasives are:

−*Paper*: Paper is used for a variety of operations from hand sanding to mechanical sanding. It is the least expensive backing. Due to the fine surface of paper, a consistent finish is produced. Paper weights include A, B, C, D, E and F weights with A being the lightest and most flexible and F being the heaviest and least flexible. A, B, C and D weight papers are used for hand sanding and light mechanical operations in the form of sheets, grip-on and stick-on discs and stick-on rolls. E and F weight papers are primarily used for more aggressive mechanical operations in the form of belts and discs (US Environmental Protection Agency, 2013).

−*Cloth*: Cloth backings used for coated abrasives are identified by weight. Cloth backings are filled or "finished" with a variety of materials, glues or resins, to create various backing characteristics, most notably flexibility.

There are three basic weights of cloth: J-weight or "jeans" is the lightest and most flexible. X-weight or "drills" is a heavier cloth that ranges in flexibility, strength and durability and is used on the broadest range of applications.

Y-weight is a heavyweight drills cloth used on heavy-duty, high stock removal operations. Several cloth types are used: cotton, polyester, and polyester/cotton blends.

−*Fiber*: Vulcanized fiber (cotton fibers which are chemically treated and then pressed under temperature and pressure to form a very durable backing) is used exclusively as the backing for resin fiber discs.

2.9. Coated Abrasive Product Manufacturing

Coated abrasives consist of sized abrasive grains held by a film of adhesive to a flexible backing. The backing may be film, cloth, paper, vulcanized fiber, or a combination of these materials.

Various types of resins, glues, and varnishes are used as adhesives or bonds. The glue is typically animal hide glue. The resins and varnishes are generally liquid phenolics or ureas, but depending on the end use of the abrasive, they may be modified to yield shorter or longer drying times, greater strength, more flexibility, or other required properties.

Figure 2.5 shows an illustration of the production process of coated abrasive products beginning with printing of a length of backing passed through a press that imprints the brand name, manufacturer, abrasive, grade number, and other identifications on the back.

Then the backing receives the first application of adhesive bond, the "make" coat, in a carefully regulated film, varying in concentration and quantity according to the particle size of the abrasive to be bonded. Next, the selected abrasive grains are applied either by a mechanical or an electrostatic method. Virtually all of the abrasive grains used for coated abrasive products are either silicon carbide or aluminum oxide, augmented by small quantities of natural garnet or emery for woodworking, and minute amounts of diamond or CBN.

In mechanical application, the abrasive grains are poured in a controlled stream onto the adhesive-impregnated backing, or the impregnated backing is passed through a tray of abrasive thereby picking up the grains. In the electrostatic method, the adhesive-impregnated backing is passed with the adhesive-coated side down over a tray of abrasive grains, while at the same time passing an electric current through the abrasive. The electrostatic charge induced by the current causes the grains to imbed upright in the wet bond on the backing. In effect the sharp cutting edges of the grain are bonded perpendicular to the backing. It also causes the individual grains to be spaced more evenly due to individual grain repulsion. The amount of abrasive grains deposited on the backing can be controlled extremely accurately by adjusting the abrasive stream and manipulating the speed of the backing sheet through the abrasive (US Environmental Protection Agency, 2013).

After the abrasive is applied, the product is carried by a festoon conveyor system through a drying chamber to the sizing unit, where a second layer of adhesive, called the size coat or sand size, is applied. The size coat unites with the make coat to anchor the abrasive grains securely. The coated material is then carried by another longer festoon conveyor through the final drying and curing chamber in which the temperature and humidity are closely controlled to ensure uniform drying and curing.

When the bond is properly dried and cured, the coated abrasive is wound into jumbo rolls and stored for subsequent conversion into marketable forms of coated abrasives as the final stage in the process shown in figure 2.5.

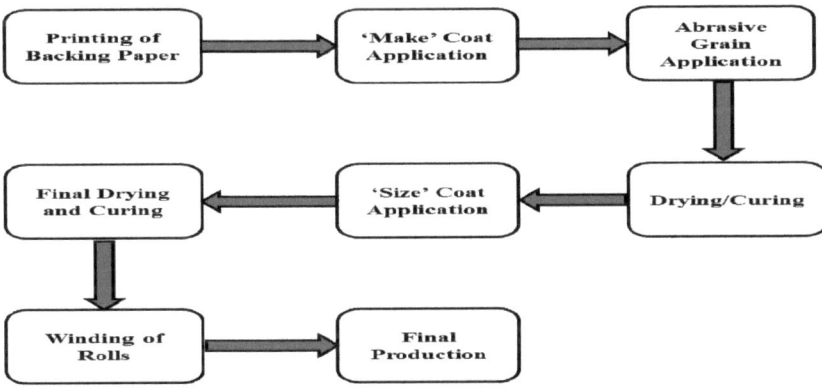

Fig 2.5: Process flow diagram for manufacturing of Coated Abrasives (US EPA)

2.10. Polyester Resins

Polyesters are an important class of high performance and engineering polymers which find use in diverse applications. Polyester is produced when dihydric alcohol like ethylene glycol reacts with an aromatic acid like phthalic acid to produce a polymeric ester. They were primarily used in compression molding (sheet molding compounds), injection molding (bulk molding compounds), resin transfer molding, pultrusion, filament winding and hand lay-up process Different kinds of polyesters were synthesized over the past decades from various types of diacid chlorides and diols (Srinivas and Bharat, 2011).

Some of the characteristics of polyester resins are high chemical and corrosion resistance, good mechanical and thermal properties, outstanding adhesion to various substrates, low shrinkage on curing, good electrical insulating properties, and the ability to be processed under a variety of conditions (Shaw, 1994). They have been used in many industrial applications such as in surface coatings, adhesives and as structural insulating materials for electronic devices (May, 1988).

2.11. Review of Other Related Works

Wai and Lilly (2002) worked on manufacturing of emery cloth/sand paper from local sourced materials. They used silicon sand (quartz) as their abrasive grits and processed it by sieving into fine grit 180µm and coarse grit 50µm. The bonds used were epoxy resins. They obtained samples of produced sand paper by adopting the hand spray method of producing sand paper, and recommended the manufacturing process for small scale industries based on successful pilot work.

Aku et al (2012) worked on characterization of periwinkle shell as asbestos-free brake pad materials using spectroscopic and wear analysis. They obtained hardness values and wear rate of periwinkle shell particles as 75 HRC and 0.2 mg/m respectively and deduced that the hardness values and wear resistance of periwinkle shell particles is higher than those of asbestos. They concluded from the results that periwinkle shells can be used for brake pad production.

Odior and Oyawale (2011) studied the formulation and manufacture of silicon carbide abrasives using locally sourced raw materials in Nigeria. The Taguchi method was used to conduct a systematic search for an optimal formulation of silicon carbide abrasives on five local raw material substitutes identified through a pilot study which were quartz, coal, sodium carbonate, saw dust and sodium chloride. The mixture was fired in a furnace to 1600°C for 6 hours forming silicon carbide chunks, which were crushed and sieved into coarse and fine grades of abrasive grains of international standard.

Dagwa et al (2012) worked on characterization of palm kernel shell powder for use in polymer matrix composites using a differential scanning calorimetry (DSC), X-ray diffraction (X-RD) and scanning electron microscopy (SEM) as well as determining the average true density, powder porosity and hydration capacity. They observed that palm kernel shell from EDX analysis has a high carbon content of about 63 wt %, with values of density, porosity and hydration capacity given as 1.58 ±0.07g/cm^3, 6.76 ±0.42%, and 150.08 ±76.91%, respectively.

Dan-Asabe et al (2012) worked on the production of composite brake pad with varied constituent compositions. The materials used were coconut shell powder, cast iron fillings, silica, epoxy (liquid resin), methyl ethyl ketone peroxide as catalyst and cobalt nephthanate as accelerator. Series of tests were conducted to ascertain composition with the best property as compared to a commercial Honda brake pad (Enuco) model widely used in Nigeria. Their results showed that higher percentage of coconut shell powder induces brittleness since compositions with lower percentage of it produced higher breaking strength and lower wear rate.

Koya and Fono (2010) developed asbestos-free automotive brake pad using palm kernel shell (PKS) as frictional filler material. They evaluated the physical, thermal, mechanical and tribological properties of the PKS-based brake pads and compared them with the values of the asbestos-based brake pads. They observed that they were better than the asbestos-based brake-pad in terms of lower specific gravity; lower percentage swelling, when wet; higher heat resistance, heat dissipation and coefficient of friction, and concluded that PKS was suitable for use as friction material in automotive brake-pads.

Ofem and Umar (2012) investigated the effect of filler content on the mechanical properties of periwinkle shell reinforced cashew nut shell liquid (CNSL) resin composites working with three particle sizes (400µm, 600µm, and 800µm), varying the filler content (10% to 40% by weight) and further characterized the composite using some mechanical tests. They found that there was an improvement on the mechanical properties as the filler content increases while these properties decrease as filler particle size increases. The optimum properties were observed at 30% filler content and 400µm particle size.

Adewuyi and Adegoke (2008) compared a total of 300 concrete cubes of size 150 × 150 × 150 mm^3 with different percentages by weight of crushed granite to periwinkle shells as coarse aggregate in the order 100:0, 75:25, 50:50, 25:75 and 0:100, which were cast, tested and their physical and mechanical properties were determined. Compressive strength from their tests showed that 35.4% and 42.5% of the periwinkle shells in replacement for granite was quite adequate with no compromise in compressive strength requirements for concrete mix ratios 1:2:4 and 1:3:6. This corresponds to savings of 14.8% and 17.5% for 1:2:4 and 1:3:6 concrete mixes, respectively.

Imoisili et al., (2012) investigated the effect of concentration of coconut shell ash on the tensile properties of epoxy composites. He produced five filler concentrations with varying weight percentages from 5 to 25 wt. %. From the mechanical tests it was found that tensile strength, elastic modulus, and micro-hardness of the composite increased with increase in filler concentration, while percentage elongation and load at fracture decreased with increase in filler concentration. He concluded that coconut shell ash can be used as reinforcing filler in epoxy resin.

2.12.   Definition of Terms

2.12.1. Periwinkle shells

Periwinkles (*Nodilittorina radiata*) are small greenish blue marine snails with spiral conical shell and round aperture (Adewuyi and Adegoke, 2008). They are common in the riverine areas and coastal regions of Nigeria where they are used for food.

2.12.2. Palm kernel shells

Palm kernel shells are hard stony endocarps that surround the kernel of a palm oil fruit. They are obtained as by-products or 'wastes' during processing and extraction of palm kernel from the nuts of the palm tree (Dagwa et al, 2012). After extraction of the palm oil the nuts are broken and the kernels are removed with the shells mostly left as waste.

2.12.3. Abrasive machining

Abrasive machining is the final stage in machining or woodwork production process that involves surface removal of workpiece parts being machined. It is an important procedure in that it imparts high dimensional accuracy and surface finish (Jain, 2008).

An abrasive is a small, hard particle having sharp edges and an irregular shape called **grit**. Grits are characterized by sharp cutting points, hardness, chemical stability and wear resistance. They are held together by a suitable bonding material to give the shape of the tool (Kalpakjian and Schmid, 2006).

2.12.4. Sand paper/ emery cloth

Sandpaper is a type of paper whose surface has been covered with an abrasive material and is used for sanding and smoothening in woodwork or metal work. They consist, basically, of a single layer of abrasive particles held to a flexible backing material by an adhesive bond. (Sandpaper, 2013). Emery cloth is described as a cloth material coated with emery powder, a type of abrasive, and used mainly for polishing metals.

CHAPTER THREE

MATERIALS AND METHODS

3.1. Introduction

This chapter describes the pilot study and experimental procedures followed for characterization and physico-chemical and mechanical evaluation used in comparing the binding effect of the polyester binder upon periwinkle and palm kernel particles at high concentrations and differing sieve sizes. Furthermore the production procedure of the emery cloth is included.

3.2. Materials/Equipment

The periwinkle (tympanotus fuscatus) and palm-kernel shells used in this work were obtained from a market in Ikot Ekpene, Akwa Ibom state and 'Sabon Gari' area of Zaria, Kaduna state respectively (plates 3.1 and 3.2).

Plate 3.1: Palm Kernel Shells (PKS)   **Plate 3.2: Periwinkle Shells (PWS)**

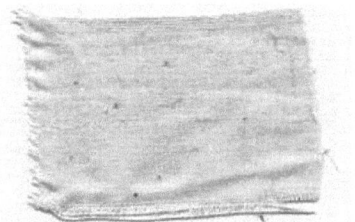

Plate 3.3: Calico fabric as backing material

Other materials used were polyester resin, methyl ethyl ketone peroxide (MEKP), cobalt nephthanate, 0.211 mm paper, calico fabric material. The resin, catalysts and hardener were purchased in hardware and chemical shops located in Zaria. The equipment used during the course of this work were: ball milling machine, set of sieves (ASTM E11), digital weighing machine, mechanical mixer, Rockwell hardness machine, universal testing machine, pin on disc machine.

3.3.   Materials Preparation

3.3.1. Preparing of periwinkle and palm kernel shells

A quantity of approximately 2 kg each of periwinkle and palm kernel shells were sun dried, followed by oven drying at 100°C for 3 hours until moisture content was reduced to the barest minimum. Further on the periwinkle and palm kernel shells were charged into a ball milling machine, milled and then sieved using three ASTM E11 sieve sizes of 105μm, 250μm and 420μm to categorize the periwinkle and palm kernel powder into FEPA abrasive grits of P140, P60 and P40 standard grits respectively.

3.3.2. Sample preparation of PWS and PKS - polyester resin composite

The digital weighing balance was used to weigh out 114g, 111.6g, 109.2g, 106.8g and 104.4g grams each of PWS and PKS powder which corresponds to 95 wt.%, 93 wt.%, 91 wt.%, 89wt.% and 87 wt.%. After weighing, they were poured into separate clean plastic containers. A double measure of polyester resin in 4.8g, 7.2g, 9.6g, 12.0g and 14.4g which corresponds to 4 wt.%, 6 wt.%, 8 wt.%, 10 wt.% and 12 wt.% was weighed and added to PWS and PKS powder respectively in their plastic containers, followed by 0.6g of cobalt naphthalene accelerator and 0.6 g of methyl ethyl ketone peroxide hardener, each making the balance of 100 wt. % in material composition, into all containers. Table 3.1 shows the adopted formulation of the composites. The mixture was blended one after the other in a mechanical mixer for 5 minutes into a thick paste.

Plate 3.5: Compacted PWS- resin samples

Plate 3.6: Compacted PKS-resin samples

Plate 3.4: Hydraulic press

Table 3.1: Batch Formulation of Composite

Materials	Weight percent of varied composition				
Periwinkle shell grains Palm kernel shell grains	95%	93%	91%	89%	87%
Polyester Resin	4%	6%	8%	10%	12%
Cobalt naphthalene accelerator	0.5%	0.5%	0.5%	0.5%	0.5%
Methyl Ethyl Ketone Peroxide Catalyst	0.5%	0.5%	0.5%	0.5%	0.5%
Total composition	100%	100%	100%	100%	100%

Then 6 test samples from each weight composition of resin were produced for the periwinkle and palm kernel shells respectively using a hydraulic press to compress the pastes to solid shapes in a metal mould of dimensions height 80mm and diameter 25mm, and die height of 65mm. Compression was carried out at room temperature at a fixed pressure of 15.5N/mm^2.

3.4. Production of Emery cloth and Sandpaper

A minimum composition mixture of 87 wt% PWS powder, 12 wt% polyester resin, 0.5 wt% for methyl ethyl ketone peroxide hardener and cobalt naphthalene each was adopted for emery cloth and sandpaper production. This was because this composition mixture had the most superior properties of hardness, wear resistance and ultimate compressive strength after conclusive testing of the various composition mixtures of PWS to resin binder. PKS was not selected as material for further emery cloth and sandpaper production because its composite mixtures showed weaker properties in the prior physico-mechanical tests carried out. The emery cloth and sandpaper was produced using periwinkle grains as abrasives and polyester resin as binder upon two different backing materials; calico fabric material and 0.2mm thickness paper. Resin, catalyst and hardener solution was prepared by stirring thoroughly using a small stick in a plastic cup. In its pre-set state, a brush was used to apply an even coat of the prepared solution upon the surface of backing material for emery cloth and sandpaper. PWS powder was poured out by hand in a gentle circular motion unto applied coat. The calico fabric and paper were cured in one hour at room temperature.

3.5. Chemical Characterization of Materials

3.5.1. X-Ray diffractometry (XRD) and X-Ray fluorescence (XRF) analysis on PWS & PKS

Equipment used for XRD of periwinkle powder was an automated Empyrean pan-analytical X-Ray diffractometer. It comprised: the source, sensor/receiver, monitor and X-ray projector. It was set at 30Kv, 10Ma. The X-ray source is a Cu LFF HR XRD tube (or Cu K α- radiation). A sample of periwinkle shell powder was added into the magazine and then loaded into the chamber of the diffractometer. The tests were carried out at the National Geological Survey Agency, Barnawa, Kaduna.

The equipment used for XRF of periwinkle and palm kernel powder was a PW 4030 X-ray spectrometer. The sample for analysis was weighed and ground in an agate mortar and a binder (PVC dissolved in toluene) was added to the sample, carefully mixed and pressed into a pellet using a hydraulic press. The pellet was loaded in the sample chamber of the spectrometer and voltage (30Kv maximum) and a current (1 mA maximum) was applied to produce the X-rays to excite the sample for 10 minutes. The spectrum from the sample was analyzed to determine the concentration of the elements in the sample. These tests were carried out at the CERT, ABU Zaria.

3.6. Physico-Mechanical Analysis of PWS and PKS - Polyester Resin Composite Samples

3.6.1. Density and percentage liquid absorption

Density measurements were carried out on the compacted PWS and PKS samples using Archimedes' principle. The buoyant force on a submerged object is equal to the weight of the fluid displaced. This principle is useful for determining the volume and therefore the density of an irregularly shaped object by measuring its mass in air and its effective mass when submerged in water (density =1 g/cc). This effective mass under water was its actual mass minus the mass of the fluid displaced. The difference between the real and effective mass therefore gives the mass of water displaced and allows the calculation of the volume of the irregularly shaped object. The mass divided by the volume thus determined gives a measure of the average density of the sample (Aigbodion et al., 2010).

The 24-h water and oil soak test determines the water and oil absorption behavior of the PWS and PKS produced composite samples and the effect of the absorbed water and oil on its dimensions. After drying of the specimens in open air for 72 hours, its weight was measured. Subsequently, the dimensions (thickness) of the specimens were measured using a vernier caliper after twenty-four hours of submersion in water and engine oil, SEA 20/50 at 26–30°C. The specimens were weighed after the excess water and oil had drained off. Eq. (1) was used in calculation of the percentage of water and oil absorption (Dagwa and Ibhadode, 2006):

$$Liquid\ Absorption\ (\%) = \frac{Wi-Wo}{Wo}\ x\ 100\%\ , \tag{1}$$

where Wi = weight after immersion and Wo = weight before immersion

3.6.2. Hardness test

The hardness values of the periwinkle shell/polyester resin composite were determined according to the American Society of Testing and Materials (ASTM E18 -79). The Rockwell hardness tester on 'B' scale (38506) with 1.56mm steel ball indenter, minor load of 10kg, major load of 100kg and standard block of hardness value 101.2HRB was used. The hardness tests were carried out in the Department of Metallurgical and Material engineering, ABU Zaria. Further comparative hardness tests were done using the Shore A durometer equipment and carried out in Nigerian Institute of Leather Science and Technology, Zaria.

3.6.3. Compressive strength test

The compressive strength test was carried out using a Norwood universal testing machine with a nominal testing force of 100kN. The diameter of test samples was 20.5mm and the cross sectional area was 331.68 mm^2. Tests were carried out in the Strength of Materials Laboratory, ABU Zaria.

3.6.4. Wear test

The wear rate of the sample was measured using a pin on disc machine (ASTM G99 - 95) by sliding it over a cast iron surface at a load of 40, 60, 80, 100 and 120kg, sliding speed of 2.4m/s and time of 20minutes. All tests were conducted at 50 and 150°C. The tests were carried out in the University of the Witwatersrand, Johannesburg, South Africa. The initial weight of the samples was measured using a single pan electronic weighing machine with an accuracy of 0.0001g. During the test, the pin was pressed against the counterpart rotating against a cast iron disc (hardness 65 HRC) of counter surface roughness of 0.3µm by applying the load. A friction detecting arm

29

connected to a strain gauge held and loaded the pin samples vertically into the rotating hardened cast iron disc. After running through a fixed sliding distance, the samples were removed, cleaned with acetone, dried, and weighed to determine the weight loss due to wear. The differences in weight measured before and after tests give the wear of the samples. The formula used to convert the weight loss into wear rate is:

$$\text{Wear rate} = \frac{\Delta W}{S} \tag{2}$$

Where ΔW is the weight difference of the sample before and after the test in mg, S is total sliding distance in m.   The coefficient of friction was then calculated by:

$$\mu = \frac{F_f}{N}, \tag{3}$$

where μ is the coefficient of friction, F_f is the frictional force read direct from the friction detecting arm strain gauge and N is the normal reaction.

3.7. Morphology of Composite Samples and Produced Emery Cloth

3.7.1. Scanning electron microscopy (SEM) of composite sample, emery cloth

Produced composite samples and final emery cloth were viewed using Phenom ProX scanning electron microscope with a magnification of >2000x. The sample for investigation was made conductive to the passage of electrons by gold spraying the sample for 5 seconds using a spouting machine. The setup was then loaded into the column which is connected to the monitor in a closed loop for which control and feedback are actualized. A finely focused electron beam with a voltage energy of 15Kv was scanned across the surface of the sample and generates secondary electrons, backscattered electrons, and X-rays. The magnification is computed by the ratio of the image width of the output medium divided by the field width of the scanned area. The tests were carried out at the Chemical Engineering Department, ABU Zaria.

CHAPTER FOUR

RESULTS AND DISCUSSION

4.1. XRD and XRF Results for PWS and PKS

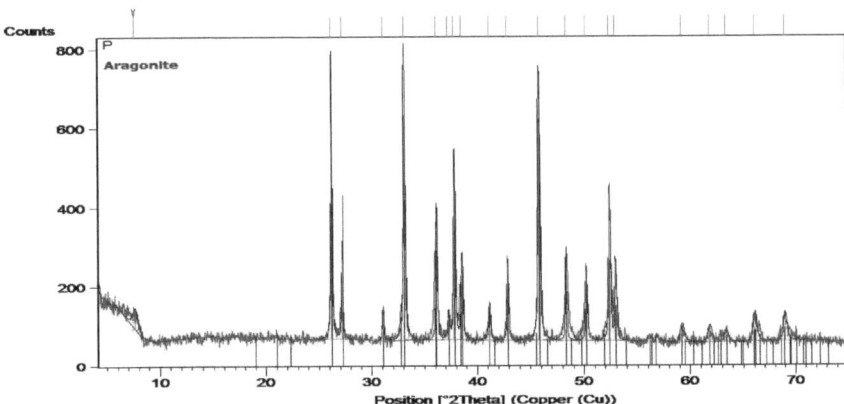

Fig 4.1. XRD Spectrum of Periwinkle Shell particles

The several sharp peaks of the XRD diffractogram in Fig 4.1 suggest that the PWS powder is mainly crystalline in form. The analysis report shows that the powder is of the orthorhombic crystal system. The mineral compound is called aragonite, with compound name calcium carbonate and formula $Ca(CO_3)$. The XRF analysis shows periwinkle shell to contain the following elements in Table 4.1.

Table 4.1: XRF analysis showing elemental composition of PWS powder

Element	Concentration
SO_3	0.33%
SiO_2	1.0%
Fe_2O_3	0.94%
CaO	93.90%
TiO_2	0.09%
MnO	0.10%
Cr_2O_3	0.05%

The XRF results above is in agreement with the XRF analysis on periwinkle shells carried out by Aku et al. (2012) on characterization of periwinkle shells as asbestos-free brake materials. Presence of hard elements like CaO, Fe_2O_3, Cr_2O_3 and SiO_2 indicates some element of hardness of the periwinkle shells. SiO_2 which is quartz is a natural abrasive, CaO is an alkaline crystal solid that occurs in natural state as limestone and known for its hardness and brittleness.

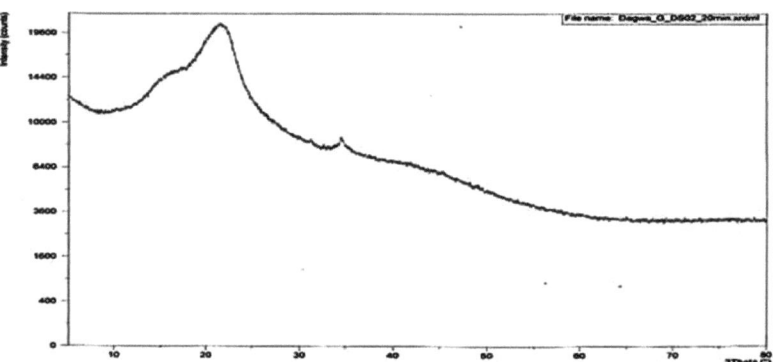

Fig 4.2: XRD Spectrum of Palm-Kernel Shells (Dagwa et al, 2012)

According to Dagwa et al (2012), the broad peak shape of the diffractogram (Figure 4.2) suggests that the PKS powder was in amorphous state while very small sharp peaks suggests small amount of micro crystalline materials could be present.

The XRF analysis of palm kernel shell results showed the presence of two metal oxides which are K_2O and CaO with 0.42% and 0.55% respectively. This indicates that the bulk of elements making up palm kernel shell powder fall into the category that cannot be detected using XRF analysis which are carbon and hydrogen.

4.2. Results of Density and Liquid Absorption for PWS and PKS-Resin Composites

From the density tests carried out as seen in Table 4.2, it is observed that for the PWS- resin composite samples, the density increases with decreasing concentration of periwinkle particles (Figure 4.3). This is as a result of the denser periwinkle particles interacting with the increasing polyester resin, and the corresponding reducing liquid absorption values indicates a reduced presence of voids and pore spaces.

For the PKS-resin composite samples however the density reduces with decreasing concentration of PKS particles due to presence of pore spaces as seen by the higher values of liquid absorption.

Table 4.2: Density and % Oil Absorption Values for PWS and PKS Composites

PWS Composite Samples @106 μm	Density	% Oil absorption @ 72 hours
95 wt% PWS + 4 wt% polyester	1.84	10.56
93 wt% PWS + 6 wt% polyester	2.26	4.62
91 wt% PWS + 8 wt% polyester	2.28	3.08
89 wt% PWS + 10 wt% polyester	2.36	1.52
87 wt% PWS + 12 wt% polyester	2.38	1.03
PKS Composite Samples @106 μm	Density	% Oil absorption @ 72 hours
95 wt% PKS + 4 wt % polyester	1.31	27.56
93 wt% PKS + 6 wt% polyester	1.24	26.53
91 wt% PKS + 8 wt% polyester	0.85	24.51
89 wt% PKS + 10 wt% polyester	1.00	28.13
87 wt% PKS + 12 wt% polyester	1.02	29.41

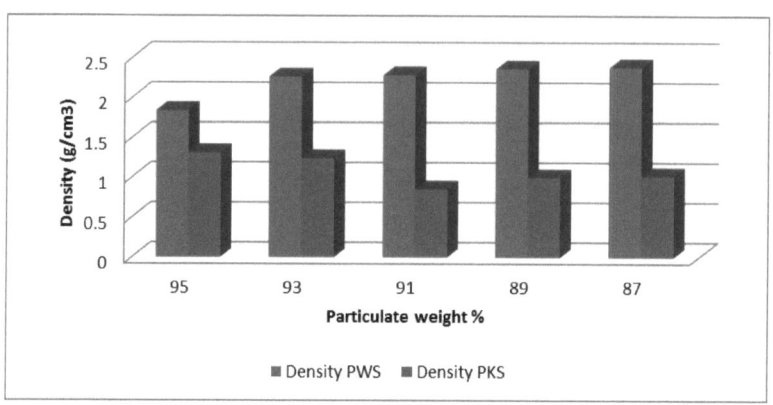

Fig 4.3: Bar chart of densities of PWS and PKS-resin composites

4.3. Result of Compressive Strength Test for PWS and PKS Composite

Table 4.3.Max Compressive Strength Results for PWS and PKS composites

Periwinkle Shell Powder @ 106µm			
Composition	Max load sustained (N)	C/S area (mm^2)	Max Comp. Strength (N/mm^2)
95 wt% PWS + 4 wt% polyester	2,770	331.68	8.35
93 wt% PWS + 6 wt% polyester	5,770	331.68	17.40
91 wt% PWS + 8 wt% polyester	8,630	331.68	26.02
89 wt% PWS + 10 wt% polyester	14,690	331.68	44.29
87 wt% PWS + 12 wt% polyester	20,730	331.68	62.50
Palm Kernel Shell Powder @106 µm			
95 wt% PKS + 4 wt % polyester	1,350	331.68	4.07
93 wt% PKS + 6 wt% polyester	1,870	331.68	5.64
91 wt% PKS + 8 wt% polyester	2,530	331.68	7.63
89 wt% PKS + 10 wt% polyester	2,540	331.68	7.66
87 wt% PKS + 12 wt% polyester	2,620	331.68	7.90

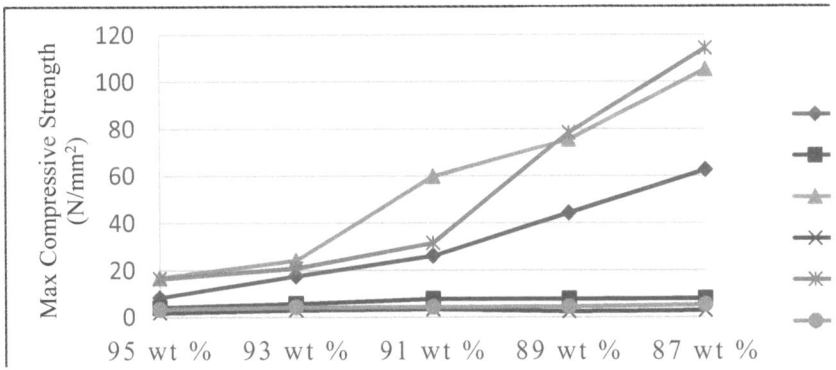

Fig 4.4: Plot of Max Compressive Strength of PWS and PKS by wt. content

The results obtained from the compressive strength tests on PWS and PKS composite samples as seen in Table 4.3 show that PWS composite samples have a comparably higher strength than PKS composite samples (Figure 4.4) and this is due to its higher density, reduced pore spaces and better bond interaction with the polymer resin matrix than the PKS composite with the polymer resin matrix. The highest compressive strength obtained is 62.50 N/mm^2 as this peak value comes from the composite mixture of 87 wt. % PWS and 12 wt. % polyester resin.

Across the sieve sizes it is observed that for the PWS composite samples, the 106μm sieve size display the lowest compressive strength when compared to the other sieve sizes (250μm & 420 μm) (Figure 4.5). This may be due to the powder fineness inducing brittleness upon the composite as seen by the reduced strength with reducing sieve sizes. However, this behavior is different with PKS which shows the highest strength with smallest sieve size (Figure 4.6).

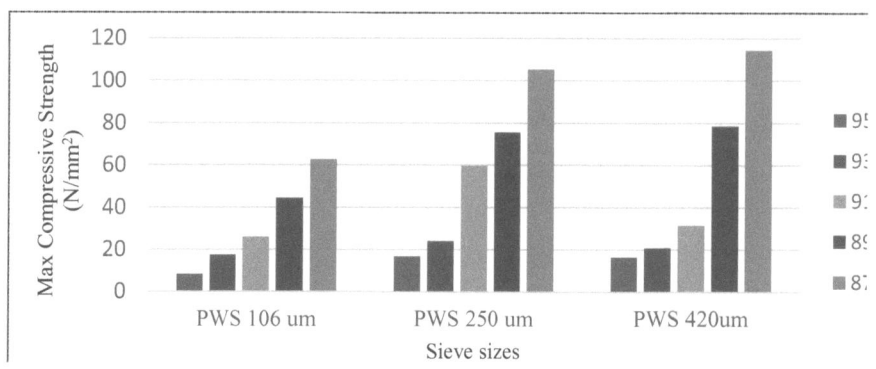

Fig 4.5: Bar chart of Max Compressive Strength of PWS by sieve sizes

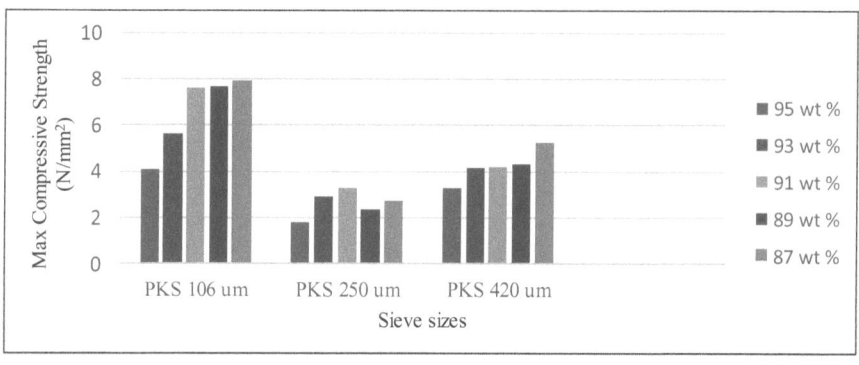

Fig 4.6: Bar chart of Max Compressive Strength of PKS by sieve sizes

4.4. Results of Hardness Value for PWS and PKS Composites

4.4.1. Rockwell hardness 'B' scale result for PWS- resin composites

Results for Rockwell hardness test done on PWS show an increase in hardness with increasing polyester concentration from 6.75 to 10.6 HRB for PWS 106μm (Figure 4.7). The increase in hardness is due to the bond interaction of the polyester which acts as a matrix holding the PWS powder particles together.

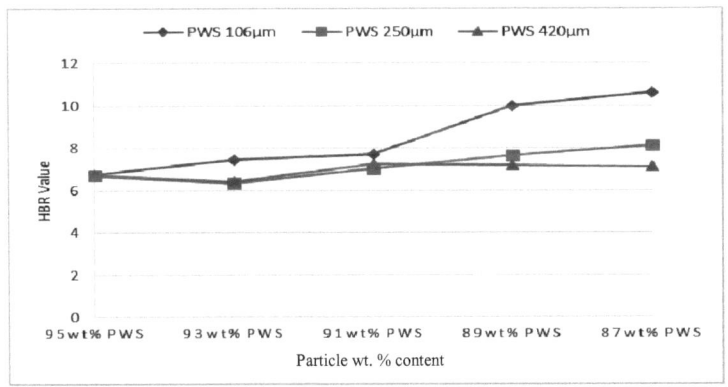

Fig 4.7: Plot of HRB value of PWS-resin composites by wt. content

Across the sieve sizes, hardness increases with reducing sieve sizes with maximum being 10.6 for PWS 106μm with 12 wt. % polyester resin (Figure 4.8)

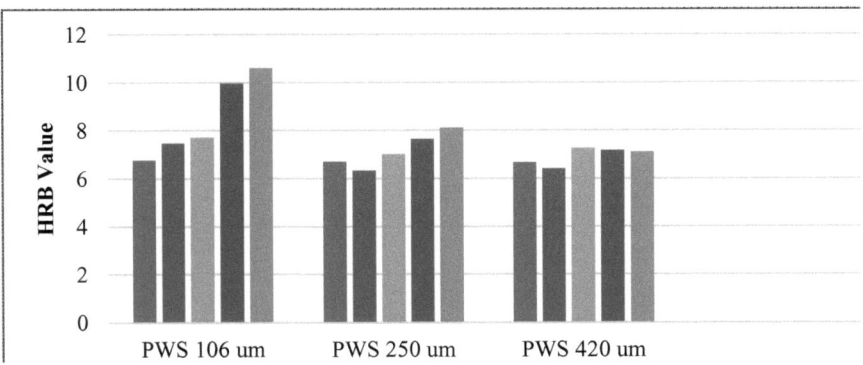

Fig 4.8: Bar chart of HRB value of PWS-resin composites by sieve sizes

4.4.2. Shore A durometer hardness comparison for PWS and PKS composites

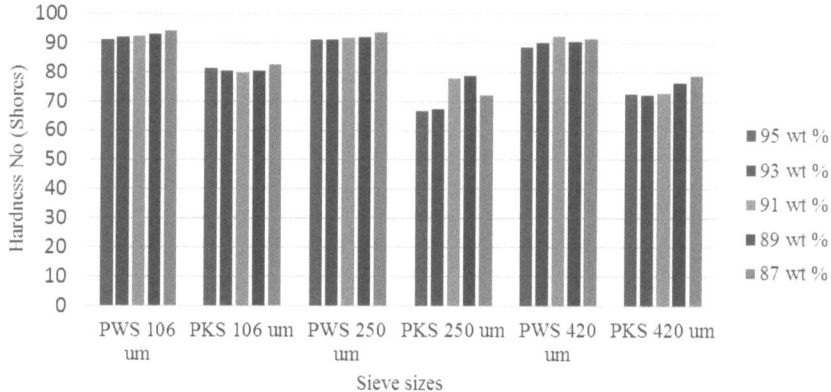

Fig 4.9: Bar chart of Shore A hardness for PWS &PKS composites

From Figure 4.9, it can be seen that for each sieve size and composition PWS-resin composite surpasses PKS-resin composite in hardness as measured with the Shore A Durometer. As seen in Figure 4.9, the highest obtained value for hardness recorded was 94.25 shores for PWS 87 wt%. The highest PKS hardness at same composition recorded was 82.75 shores.

4.5. Wear Rate for PWS-Resin Composites

Wear rate on the periwinkle shell composites was conducted at two temperatures to understand its wear behavior under heat generated by friction during service conditions as emery cloths and sandpapers. A single grain size of 106 microns at different concentrations of the periwinkle shell particulates was tested and from the results it was observed that the wear rate increases with increasing periwinkle shell particles concentration, applied load and temperature (Figure 4.10). The graph of coefficient of friction as seen in Fig 4.10 shows that with increasing load there is a reduction in the frictional coefficient of PWS composite samples against the counterpart rotating hardened cast iron disc.

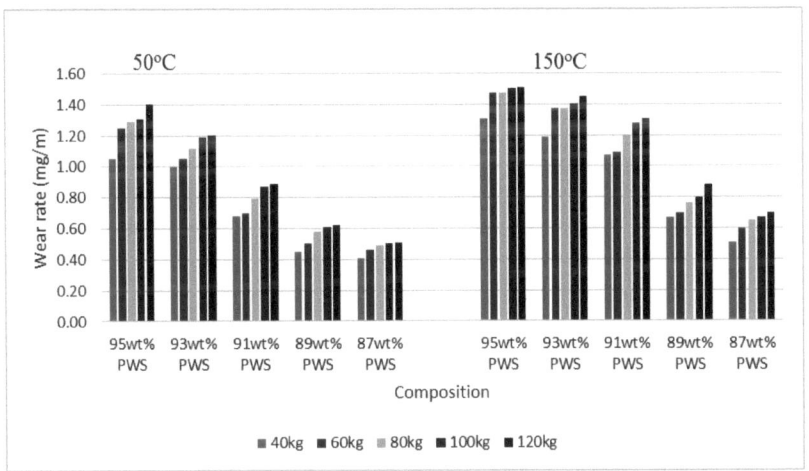

Fig 4.10: Bar chart for Wear Rate of PWS 106µm at 50°C and 150°C

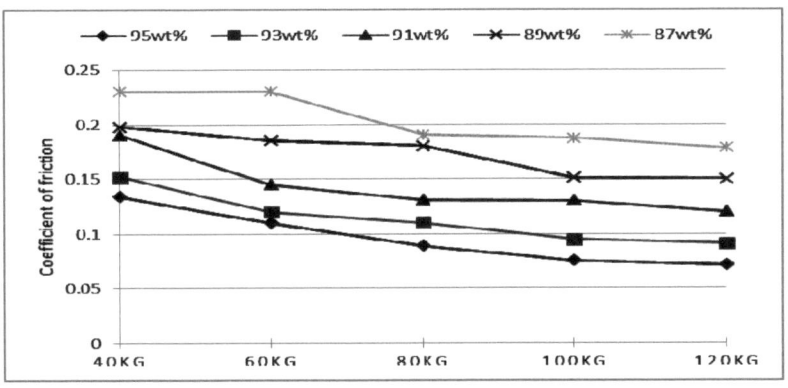

Fig 4.11: Plot of Coefficient of Friction for PWS 106µm at 50°C

4.6.   SEM Imaging/Morphology of PWS and PKS- Resin Composites

4.6.1. Morphology for PKS-resin composites

Plate 4.1: PKS at 93 wt. % (3800x mag.)   **Plate 4.2: PKS at 87 wt. % (2400x mag.)**

Plate 4.1 is microstructure of PKS/resin composite at 93 wt. % PKS. An effect of compressive forces on PKS grains is clearly seen. Plate 4.2 microstructure images of PKS 250µm particles at 87 wt. % polymer matrix composite show more broken up particles bonded with the polyester resin than images of PKS 250µm particles at 93 wt. % as seen in Plate 4.1. Void spaces between the bonded particles in the microstructure can be seen in the Plate 4.2.

4.6.2. Morphology of PWS-resin composites

Plate 4.3: PWS at 95 wt. % (1050x mag.)   **Plate 4.4: PWS at 87 wt. % (2250x mag.)**

SEM imaging of microstructure seen in Plate 4.3 is the 95 wt. % PWS 250μm particle size polymer matrix. The image show grains at disarray and out of shape, which reflects brittleness due to the low concentration of resin causing particles to become distorted under compressive forces. In Plate 4.4 the microstructure show the PWS grains held in position by the polymer matrix with more longitudinal shaped aligned grains than image in Plate 4.3. This indicates improved interfacial bonding as a result of increase in polyester resin content. Also the resin holds the grains in position against applied compressive forces and there is less distortional effect on grains.

Images in plate 4.5 of EDS scan of the PWS composite show the presence the elements calcium, carbon, silicon and antimony in various weight compositions of 59.8, 18.5, 16.2 and 5.6 respectively. This validates the XRF analysis on periwinkle shell having calcium and carbon as predominantly present in compound state of $CaCO_3$ (see Table 4.1). EDS of PKS composite shows presence of silicon, nitrogen, oxygen and carbon in wt% of 1.6, 8.4, 76.9 and 13.1 respectively as seen in Plate 4.6. The high presence of oxygen detected can be attributed to large presence of pore spaces which traps air in the PKS composite samples.

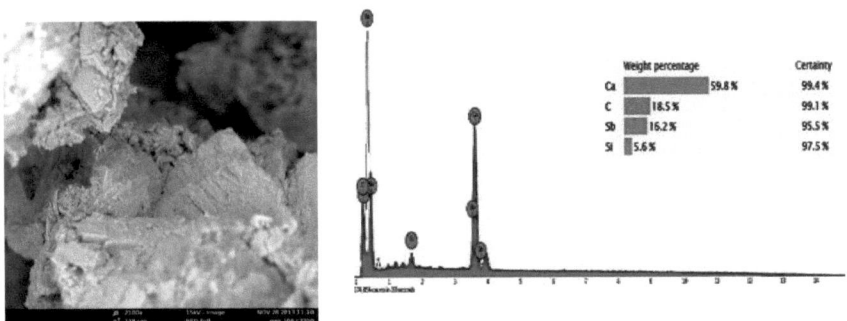

Plate 4.5: 2100x mag PWS 106μm at 91 wt. % with EDS profile

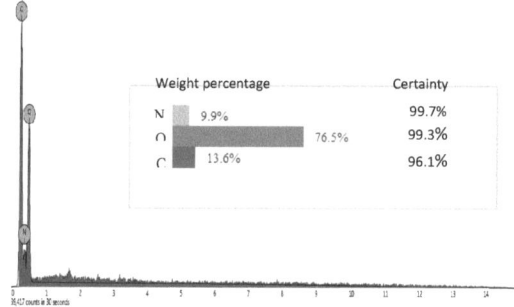

Weight percentage		Certainty
N	9.9%	99.7%
O	76.5%	99.3%
C	13.6%	96.1%

Plate 4.6: 2200x mag PKS 106µm at 91 wt. % with EDS profile

4.7. Optimization Analysis of Materials and Compositions for Production of Emery Cloth

From the physico-chemical analysis, mechanical and SEM tests carried out on composite mixture of PWS and PKS with polyester resin binder, the PWS grains showed greater prospects as abrasives than when compared to PKS grains. This can be seen in Table 4.4

Table 4.4: Comparison of Properties between PWS and PKS-Resin Composites

Properties	87wt. % PWS composite	87 wt. % PKS composite
Density	2.38 g/cm^3	1.02 g/cm^3
Oil Adsorption	1.03%	29.41%
Compressive strength	62.50 N/mm^2	7.90 N/mm^2
Hardness Shores	94.25 shores	82.75 shores
Hardness HRB	10.6 HRB	Too soft to be detected
Crystal structure	Mainly crystalline	Mainly amorphous
Elements in order of wt. %	59.8% Ca, 18.5% C, 16.2% Si, 8.7% Sb	76.9% O$_2$, 13.1% C, 8.4% Ni, 1.4% Si
SEM Morphology	Good interfacial bonding, more compact structure	Uneven bonding caused by air enclosed pore spaces.
Summary	**Hard grains**	**Soft grains**

Therefore periwinkle grains were selected for purpose of production of emery cloth. Considering the composition of PWS to resin, a comparison of the properties is given in Table 4.5.

Table 4.5: Comparison of Properties between Compositions of PWS Composites

Composition / Properties	95wt% PWS, 4wt% resin	93wt% PWS, 6wt% resin	91wt% PWS, 8wt% resin	89wt% PWS 10wt% resin	87wt% PWS, 12wt% resin
Strength	8.35 N/mm^2	17.4 N/mm^2	26.02 N/mm^2	44.29 N/mm^2	62.50 N/mm^2
Hardness	6.75 HRB	7.45 HRB	7.70 HRB	9.97 HRB	10.6 HRB
Wear rate @40kg, 50°C	1.05 mg/m	1.00 mg/m	0.68 mg/m	0.45 mg/m	0.41 mg/m
Coeff. of friction @40kg, 50°C	0.134	0.152	0.19	0.198	0.23

Therefore the PWS composite of 87 wt. % particles and 12 wt. % polyester resin was used as the minimum value of composition in producing emery cloth.

4.8. SEM Imaging/Morphology of Produced Emery Cloth P40 Grade

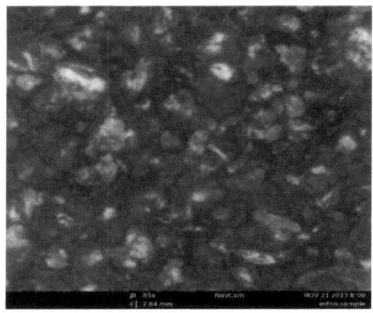

Plate 4.7: 85x mag P40 grade Emery

Plate 4.8: 2550x mag P40 grade

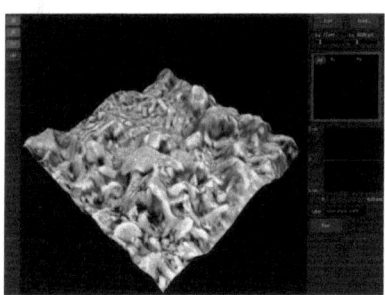

Plate 4.9: 3D Scan image

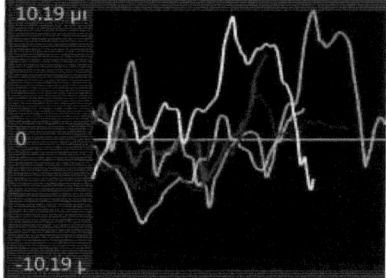

Plate 4.10: Roughness Index chart

It is observed from the surface morphology of produced PWS P40 grade emery cloth that periwinkle shell grains appear to have sharp, angular edges (plate 4.8) essential for abrasive grits, and the 3D scan image shows rough contour surfaces (plate 4.9). The degree of roughness of the surface is indicated in the roughness index chart shown in plate 4.10. The computed roughness index value from the chart from profiles one to five is given as 4.716μm and 1.63μm for indices Ra and Rz respectively.

CHAPTER FIVE

CONCLUSIONS AND RECOMMENDATIONS

5.1. Conclusions

From the results of the investigations and discussions, the following conclusions have been made.

1. That periwinkle shell grains are mainly crystalline and it is composed of over 90% calcium oxide (CaO) while palm kernel shell grains are mainly amorphous in nature and is composed of carbon, hydrogen and trace quantities of K_2O and CaO.

2. That periwinkle and palm kernel shell grains have been successfully processed and sieved to FEPA particle standards of P140, P60 and P40 grit sizes. Out of the two, periwinkle shell grains have a defining shape of sharp edges, and have been proven to abrade the surface of aluminium metal.

3. There is a gradual increase in hardness (3.5% for PWS, 1.53% for PKS/resin composite), compressive strength (648.5% for PWS, 94.1% for PKS/resin composite) and wear resistance (61% for PWS/resin composite) with increase of polyester resin content from 4 to 12 weight percent.

4. Periwinkle shell-resin composites have superior physico-mechanical properties such as density with 80% difference, oil absorption with 93.2% difference, hardness and compressive strength with 13% and 155% difference respectively, and interfacial bonding over the palm kernel shell-resin composites.

5. That the material and composition with the best properties of hardness, wear resistance and compressive strength is periwinkle shell grains at 87 wt. % particles, 12 wt. % polyester resin, and 0.5 wt. % for methyl ethyl ketone peroxide hardener and cobalt naphthalene and this was adopted for emery cloth production.

6. The Ra roughness parameter of produced PWS emery cloth was found to be 4.716μm.

5.2. Contributions to Knowledge

The followings are some of the contributions to knowledge for carrying out this present research:

1. Periwinkle shell grains have been considered as a substitute for conventional abrasive grits in emery cloth and sandpaper production.

2. This research has exposed another area of application (abrasive properties) of periwinkle shells in addition to the currently established applications and thus increased the economic value of periwinkle shells.

3. From the research it has been proven that high concentrations of palm kernel grains at minimum levels (4 to 12 wt. %) of polyester binder is found to be very porous and therefore unfit for load bearing applications.

4. With further improvement on the abrasiveness of developed emery cloth and sandpaper prototype, this research will bring about reduction in the importation of foreign made sandpapers and abrasive cloth and bring foreign exchange earnings for the country.

REFERENCES

Adewuyi A. P. and Adegoke T. (2008): Exploratory Study of Periwinkle Shells as Coarse Aggregates in Concrete Works, ARPN Journal of Engineering and Applied Sciences, Vol 3(6): 1-5

Agoha, E.E.C (2007): Biomaterials from Periwinkle Composition and Functional Properties presented at the World Congress on Medical Physics and Biomedical Engineering, Aug 27 to Sept 1. COEX Seoul, Korea

Aigbodion V. S., Akadike U., Hassan S. B., Asuke F. and Agunsoye J. O. (2010): Development of Asbestos - Free Brake Pad Using Bagasse, Tribology in industry, Vol 32 (1): 12 – 18.

Aku, S.Y., Yawas, D.S., Madakson, P.B. and Amaren, S.G. (2012): Characterization of Periwinkle Shell as Asbestos-Free Brake Pad Materials. Pacific Journal of Science and Technology. 13(2): 57-63.

Badmus, M.A.O., Audu, T.O.K and Anyanta, B.U (2007): Removal of Lead Iron from industrial waste waters by activated carbon prepared from periwinkle shells (Typanotonus fuscatus), Turk. J. Eng. Environ. Sci., 31: 251 -263.

Benjamin, B. O (1989): Investigation into the use of periwinkle shells as light weight aggregate for concrete construction. MSc Project submitted to Ahmadu Bello Univeristy, Zaria.

Bob-Manuel F.G (2012): A Preliminary Study on the Population Estimation of the Periwinkles Tympanotonus Fuscatus (Linnaeus, 1758) and Pachymelania Aurita (Muller) at the Rumuolumeni Mangrove Swamp Creek, Niger Delta, Nigeria, Agriculture and Biology Journal of North America, Vol 3(6): 265-270.

Dagwa, I.M., Builders, P.F. and Achebo, J. (2012): Characterization of Palm Kernel Shell Powder for use in Polymer Matrix Composites. International Journal of

Mechanical & Mechatronics Engineering IJMME-IJENS Vol 12 (04): 88 – 93.

Dan-Asabe B., Madakson P. B and Manji J (2012): Material Selection and Production of a Cold-Worked Composite Brake Pad, World J of Engineering and Pure and Applied Sci. Vol 2 (3): 92- 97

Egonmwan, R.T. (1980). On the biology of Tympanotonus fuscatus var rodula (Gasstropoda: Prosobranchia: potamididae). Proceedings of the 8th International Malacological congress, Budapest.

Ibhadode, A. O. A. and Dagwa, I. M (2008): Development of Asbestos-Free Friction Lining Material from Palm Kernel Shell. J of the Braz. Soc. of Mech. Sci.& Eng. (02):166-173

Imoisili P. E., Ibegbulam C. M. and Adejugbe T. I. (2012): Effect of Concentration of Coconut Shell Ash on the Tensile Properties of Epoxy Composites, The Pacific Journal of Science and Technology, Vol 13(1):463 -468

Jain, R.K (2008): Production Technology, 16th Edition, Khanna Publishers, India. Pages 1011 -1045

Kalpakjian, S. and Schmid, S. (2006): Manufacturing Engineering and Technology, 5th Edition, Pearson Prentice Hall, New Jersey. Pages 790 -831

Koya, O.A., and Fono, T.R (2010). Palm kernel shell in the manufacture of automotive brake pad (accessed at www.rmrdctechnoexpo.com, May 22, 2013).

King, R.I. and Hahn, R.S. (1986): Handbook of Modern Grinding Technology, Chapman and Hall.

Malkin, S. and Guo, C. (2008), Grinding Technology: Theory and Applications of Machining With Abrasives, 2nd Edition, Industrial Press, Inc., New York.

Malu, S.P. and Bassey, G.A (2003): Periwinkle shell as alternative source of lime for glass industry, Global J. Pure and Applied Sciences, Vol 9(4): 491- 494.

Marinescu I.D, Rowe W.B, Dimitrow B, and Inasaki I (2004) Tribology of abrasive machining process, 1st Edition, William Andrew publishing, Page 268

May, C. A. (1988): Epoxy Resins Chemistry and Technology, 2nd ed., Marcel Dekker, Inc.:New York.

National Orientation and Public Affairs (1999): "Nigerian Economic Policy (1999-2003)"- Federal Ministry of Information and National Orientation, Abuja.

Njoku R. E., Okon A. E. and Ikpaki T. C. (2011): Effects of Variation of Particle Size and Weight Fraction on the Tensile Strength and Modulus of Periwinkle Shell Reinforced Polyester Composite, Nigerian Journal of Technology, Vol 30 (2).

Nwokolo, T. U. (1994). Investigation Study on Palm Kernel Shells as Light Weight Aggregate. Final Year Project submitted to Civil Engineering Department, University of Benin, Benin.

Odior A. O and Oyawale F. A. (2011): Formulation of Silicon Carbide Abrasives from Locally Sourced Raw Materials in Nigeria, Journal of World Congress on Engineering, Vol 1.

Ofem M. I. and Umar M. (2012): Effect of Filler Content on the Mechanical Properties of Periwinkle Shell Reinforced CNSL Resin Composites, ARPN Journal of Engineering and Applied Sciences, Vol 7 (2) 212- 215

Ohimain E. I., Bassey S. and Bawo D. D. S (2009): Uses of Sea Shells for Civil Construction Works in Coastal Bayelsa State, Nigeria: A Waste Management Perspective, Research Journal of Biological Sciences, Vol 4 (9): 1025 -1031

Olutoge F. A., Okeyinka O. M. and Olatunji S. O. (2012): Assesment of the Suitability of Periwinkle Shell Ash as Partial Replacement for Ordinary Portland cement in Concrete, IJRRAS Vol 10 (3) 428 -434.

Omobowale M.O (2010), Problems facing local manufacturers in the Nigerian Agro-allied machine fabrication industry, ATDF Journal, Volume 7, Issue 3/4 2010

Osarenmwinda, J.O. and Awaro, A.O (2009): The potential use of periwinkle shell as coarse aggregate for concrete. Advance Material Research, Vol 62(64): 39 - 43.

Pruti F. (2011): Development of Composite Wheels for Hard and Soft Metals, University of East London. Pages 23 -39

Shaw, S. J. (1994): Rubber Toughened Engineering Plastics, Collyer, A. A., ed., Chapman &Hall: London. Pages 165-209.

Sandpaper. (2013): In *Encyclopædia Britannica*. Retrieved March 22, 2013 from http://www.britannica.com/EBchecked/topic/522171/sandpaper

Srinivas, C.V., Bharat, K.N.,J. (2011) Material Environmental. Science 2, 351.

UGWC (1992): The Grinding Data Book, Universal Grinding Wheel Company Limited.

US Environmental Protection Agency: Abrasive Manufacturing. Retrieved March 24, 2013 from http://www.epa.gov/ttnchie1/ap42/ch11/final/c11s31.pdf

Usman, N. D., Idusuyi, F. I., Ojo, E. B. and Simon, B (2012): The Use of Sawdust and Palm kernel Shell as Substitute for Fine and Coarse Aggregates in Concrete Construction in Developing Countries. Journal of Chemical, Mechanical and Engineering Practice, Vol. 2 [3] 51-62

Wai, J.J and Lilly, M. T (2002): Manufacturing of Emery Cloth (Sand Paper) from Local Raw Materials. Global Journal of Engineering Research, Vol 1 [1] 31-37

Yamaguchi K., Wei, Y. and Takeuchi M. (1999): Development of DLC Fiber Grinding Wheel. Proceedings of the Vernal Meeting of the JSPE, Tokyo, Page 260.

APPENDIX

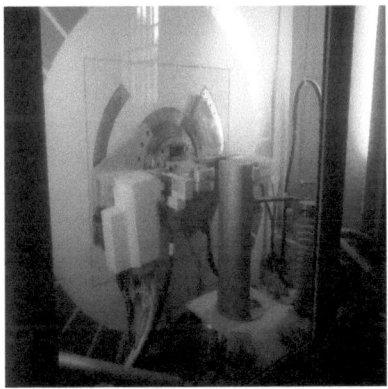

Plate B1- XRD Machine

Plate B2- Wear Test Machine

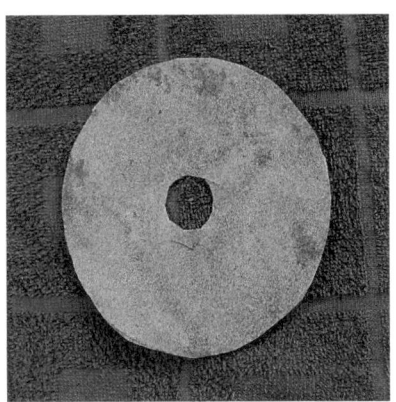

Plate B3- Produced Emery cloth (P40 grade)

Plate B4- Produced Emery cloth (P140 grade)

Plate B5- Produced P40 and P140 sandpaper samples